W9-AYO-251

Building a House

by Byron Barton

Greenwillow Books, New York

Copyright © 1981 by Byron Barton • All rights reserved. No part of this book may be reproduced in any form or by any means without permission in writing from the Publisher, Greenwillow Books, 105 Madison Avenue, New York, N.Y. 10016. Printed in U.S.A. First Edition. 8 7 6 5 4 3 2 1 Library of Congress Cataloging in Publication Data: Barton, Byron. Building a house. Summary: Briefly describes the steps in building a house. 1. House construction–Juvenile literature. [1. House construction] I. Title. TH4811.5.B37 690′.8373 80-22674 ISBN 0-688-80291-5 ISBN 0-688-84291-7 (lib. bdg.)

On a green hill

a bulldozer digs a big hole.

Builders hammer and saw.

A cement mixer pours cement.

Bricklayers lay large white blocks.

Carpenters come and make a wooden floor.

They put up walls.

They build a roof.

A bricklayer builds a fireplace and a chimney too.

A plumber puts in pipes for water.

An electrician wires for electric lights.

Carpenters put in windows and doors.

Painters paint inside and out.

The workers leave.

The house is built.

The family moves inside.